WARLORD ARMIES

Text by Tim Newark
Color plates by Angus McBride
Richard Hook, Graham Sumner
and Graham Turner

by CONCORD PUBLICATIONS CO.
603-609 Castle Peak Road
Kong Nam Industrial Building
10/F, B1, Tsuen Wan
New Territories, Hong Kong
www.concord-publications.com

We welcome authors who can help expand our range of books. If you would like to submit material, please feel free to contact us.

We are always on the look-out for new, unpublished photos for this series. If you have photos or slides or information you feel may be useful to future volumes, please send them to us for possible future publication. Full photo credits will be given upon publication.

ISBN 962-361-086-6
printed in Hong Kong

Picture Credit:
Thanks to Angus McBride, Richard Hook, Graham Sumner, Graham Turner, Linden Artists, Military Illustrated, and the Carroll Institute.

Introduction

The warlord has never gone away. Civilized armies, raised by cities or empires, organized and uniformed, have dominated past periods of warfare, but the warlord is always present in less civilized arenas. The warlord is the cult of the personality turned into a military force. In an age of chaos, he appears as the only man strong enough to pull together an army that can impose its will on other people. Even at the peak of ancient civilization in Europe, the Roman Empire could be torn apart by warlords—under the guise of being senators or generals—who chose to carve out their own realm within the boundaries of empire. Such warlords eventually broke the empire in two—East and West—and then mingling with barbarian forces finally saw it overthrown.

The currency of the warlord is fear and violence. He ignores the law of the land to impose his own version of justice. He submits to no one else. Once compromise enters the equation he ceases to be a warlord and becomes merely a hired-hand, a general.

The nature of the warlord characterizes his method of warfare. His power depends on relationships and his army functions on the battlefield according to this. A warlord's powerful friends and allies bring their own retinues of troops and so on down. They fight as a gang would fight, with a much feared and revered leader in the center of the action surrounded by other troops whose job it is to protect the leader and further his intentions. Thus, in the world of medieval warfare, it is rare to find serried rows of one type of warrior all standing together, as you might find on the table of an inexperienced wargamer. Instead, a warlord's army would be composed of clusters of troops, say armored horsemen surrounding their leader, which in turn would be supported by light cavalry which in turn are supported by archers and foot soldiers. They move and perform in mutually-supporting groups, tied together by personal loyalty and ambition, not differentiated units being guided by an invisible hand. Their method of warfare is one of opportunity ruthlessly pursued.

The warlord's preferred medium is chaos. When central control fails, they are happy to assume dominance of regions and provinces. When a state government is strong, warlords are hunted down and executed as criminals, unless they can be recruited as generals. In this book, the world of the warlord is examined from prehistory through to the 20th century, demonstrating its eternal presence.

Some of the world's finest military illustrators are assembled on the following pages. Angus McBride is renowned for his authentic and dramatic recreation of the ancient and medieval worlds. Based in South Africa, his work is now collected around the globe. Richard Hook has illustrated numerous articles and books for *Military Illustrated* and *Osprey*. He has a particular passion for recreating the world of the Native American. Graham Sumner is a Roman re-enactor with the celebrated *Ermine Street Guard* group and brings a detailed archaeological knowledge to his artwork. Graham Turner is the son of the respected aviation artist, Michael Turner, but his own specialized interest lies in medieval warfare.

Bronze Age Warlord, Southern England, 1500 years BC

[Plate 1]

The Bronze Age saw a revolution in ancient warfare. It covers a period of roughly two thousand years from 3000 BC to 1000 BC but this varies from region to region and in northern Europe the dates fall later. Before this period men fought with daggers made of flint or arrows tipped with bone. The technical achievement of making the metal bronze from an alloy of copper and tin took the ancient arms industry up a considerable notch. Swords with fine edges and tips could be made, alongside spears with sharper heads. But, of course, this didn't mean that everyone fought with such weapons. Stone maces and fire-hardened wooden spears were still used alongside the new bronze weaponry as only the wealthier warriors in a society could afford to be equipped with such splendid items.

In Britain, a complex society was already in existence during the Bronze Age. The great agricultural revolution that first occurred around 10,000 years ago meant that farming was the predominant engine of the economy. The clearing of forests continued, perhaps with greater speed thanks to bronze axes. Farmers and traders brought their goods to markets and these formed the basis of towns, some of which were protected by wooden palisades. Hill forts had begun to appear. Boats were a main form of transport and raiders could use them to attack neighboring villages sited along rivers. Small horse, like Shetland ponies, were used to pull chariots, but these were used more as impressive methods of arrival at a battlefield, when the leading warriors would then dismount and fight on foot.

The painting opposite, by Graham Sumner, reconstructs a Bronze Age warrior from 1500 years BC. He carries a stone mace, a gold breastplate and both a copper axe and dagger. The figure is based upon a body excavated in southern England on Salisbury Plain from a site called "Bush Barrow". It comes from a richly furnished burial mound and must be that of a Bronze Age warlord. The burial site is close to Stonehenge and some experts have speculated that such a powerful man may well have been involved in this most famous building project of ancient Britain.

The copper bladed dagger has a pommel decorated with minute lengths of inlaid gold thread and must have been a very expensive item. Traces of organic material found with it suggest it was carried inside a scabbard made of wood and sheepskin. The lozenge-shaped piece of sheet gold covering the warrior's chest has been called a breast-plate but clearly it was not for use in combat but more as a symbol of great wealth and status. The axe and mace have little decoration and may have been less ceremonial and more lethal.

The presence of gold in this burial site suggests trading links with Ireland, while similar shaped daggers have been found in northern France. Amber beads and cups found in other English burials indicate links with the Baltic. Wine and oil would have come up from the Mediterranean. It is no surprise to find such a complex trading network in place in this period. Long-range trade routes were vital for supplying the prestige goods demanded by warlords. Boats sailed from coast to coast and along rivers. Wheeled wagons brought goods across land. Forts were built to protect important markets and trade routes.

It would have been the international reputation of southern England's wealthiest figures that would have attracted raiders from abroad. Among them were the Celts who sailed over from France to set up their own distinctive culture on British soil. That such prehistoric conflicts were ruthless in the extreme is proved by a discovery in southern Germany in which 34 skeletons were found among the buried ruins of a village. The remains were of all ages and both sexes—all had been murdered. The only survivors would have been those useful as slaves. Slave hunting and trading was a major part of the early European economy. Warlords such as the figure reconstructed in the painting opposite would have led the resistance against the early Celtic invaders.

Pyrrhus, Greek Warlord, Macedonia, early 3rd Century BC

[Plate 2]

Alexander the Great dramatically changed warfare in ancient Greece and the Middle East. When he died in 323 BC, he left a vast empire controlling territory from his homeland in Macedonia in the west to Persia in the east. But he had not secured a dynasty to rule these lands and his leading generals—dubbed the Successors—fell out and fought each other to maintain control over their parts of his empire. The great area of conquest broke up into regions of command and the subsequent decades saw many wars fought amongst the warlords.

Pyrrhus, King of Epirus in north-west Greece, was one of these adventuring warlords. He fought several campaigns against Macedonia and then was asked to help the Greek cities of southern Italy and Sicily defend themselves against the expanding power of Rome. In 279 BC, he confronted the Romans in the most famous episode of his life. He defeated the Romans in a ferocious battle at Asculum in Apulia but his army received such heavy losses that it became a byword for a victory won at a high cost—a Pyrrhic victory. Four years later, Pyrrhus was forced out of Italy and returned to Greece where he died in a street-fight in Argos.

The painting opposite, by Angus McBride, reconstructs an earlier event in the life of Pyrrhus. In 288 BC, Pyrrhus struck a secret agreement with Lysimachus of Thrace to partition Macedonia, then under the rule of Demetrius Poliorketes. Lysimachus was one of those senior warriors who had served with Alexander the Great and his great reputation was as effective as a mighty weapon in battle. When Lysimachus invaded from the east, Demetrius took his Macedonian army to fight him, but once his soldiers learned of the great warlord's experience with Alexander they declined to face him and marched back to the west to face the then less experienced Pyrrhus.

Demetrius camped his troops before Pyrrhus, but before battle could be engaged, his less then loyal soldiers decided to change sides and joined the popular Pyrrhus. They cut down oak branches to make wreaths of oak leaves—the symbol of Epirus. The painting shows Macedonian troops of the elite unit known as the "White Shields". They fought with long spears called *sarisa* in a dense phalanx formation.

Pyrrhus is shown on horseback wearing a crested helmet decorated with goat horns. This is typical of Macedonian warlords who liked to associate themselves with Greek gods. When Alexander chose to have his image immortalized on coins, he was portrayed with horns, the sign of the god Dionysus. Dionysus was the wild god, the god of the impetuous violent spirit that lurks inside men and women and is liberated by alcohol. The horned image of Dionysus would later form the basis for Christian visions of the devil. As a son of Zeus, in the Greek myths, Dionysus left his Greek homeland to roam across the world accompanied by a wild army of satyrs and maenads, spreading the knowledge of wine, conquering armies in Egypt and Asia, and marching as far as India. He was a model for Alexander's remarkable conquests and a source of inspiration for the careers of the later Successors.

Like Alexander, Pyrrhus believed he had divine power and used his powers to heal people. According to Plutarch, his right toe was believed to possess particular divinity and when he was burned after his death, this was found to be untouched by fire.

Despite the professional aspect of warfare at this time with established formations of cavalry and infantry, there still appears to have been room for an older, more heroic form of combat. Plutarch tells a story in which Pyrrhus, who also believed himself to be descended from the Greek hero Achilles, once challenged a general of Demetrius—called Pantauchus—to single combat. At first, they threw javelins at each other but then they came to close combat with swords. Pyrrhus received one wound, but he slashed back at Pantauchus, wounding him on the thigh and neck. Finally, he pushed Pantauchus off his feet and would have killed him had not his friends come to his rescue. The personal victory inspired his troops to charge the Macedonian phalanx and break it, thus gaining victory.

Angus McBride

Roman Auxiliary on Campaign, Northern Europe, 1st Century AD

[Plate 3]

The Roman Empire was not immune to the machinations of warlords. Although its highly centralized government based in Rome created the image of a unified empire, its provinces were inevitably dominated by strong men who might in time defy the authority of the emperor. Civil war would result and Roman soldiers would be pitched against Roman soldiers. In such a situation, it would not be so much the method of warfare that proved the deciding factor, but the character of the soldiers themselves and from where they were recruited. Germanic and Gallic auxiliaries soon earned themselves a high reputation for their ferocity in battle.

The painting opposite, by Graham Sumner, recreates a Roman auxiliary of the 1st century AD recruited from northern Europe. He is more typical of the appearance of a Roman soldier on campaign. He wears a waterproofed leather cover over his mail shirt. Leather was used a lot by the Romans as a protective covering, especially for their shields. On one occasion, Caesar records that his troops were caught in a surprise attack and not able to remove their leather shield covers. Moulded leather may even have been used as a cheap form of body armor. Following the fashion of gladiators, the recreated figure wears the additional protection of cloth wrapped around his sword arm and legs. He carries an unusual flat rectangular shield—its markings indicate that the auxiliary is from the Batavi tribe that lived along the North Sea coast north of Gaul.

The most famous Roman armor, seen in almost all reconstructions and visible on many remaining stone reliefs, was the *lorica segmentata*. It was made of strips of iron plates strapped around the arms and chest. It was a purely Roman invention and is thought, perhaps, to derive from the armor worn by gladiators in the arena—suited to protection against sword blows in close combat. Such armor was worn from the 1st to 3rd centuries AD, but it was not the most typical armor worn by Roman warriors. The majority of them wore shirts of mail, composed of interlocking iron rings, which originated from the Celts.

Mail armor was popular because it was easy to wear and fight in. Modern reconstructions show that a mail shirt follows the shape of the body. This allows the wearer to take deeper breaths and wear it for longer without exhaustion, as compared to plate armor. It can also be self-cleaning, in that mail rings rub against each other, removing rust and dirt with friction. The main weakness of mail is that it is vulnerable to piercing weapons such as arrows. Thick padding may have been worn beneath the mail to counter this.

Because the Roman army has a reputation for order and discipline, it is presumed that Roman soldiers all wore the same armor and fought with the same weapons. But this was not true. Campaigning always reduces the best intentions of soldiers, ruining uniforms and putting great pressure on supplies, and the Roman army was no different. Both mail and plate armor was worn side by side, as many soldiers inherited or bought armor from previous generations of soldiers.

Different nationalities served in the Roman imperial army and they frequently brought their own local styles of dress and fighting with them. This might include warriors in Egypt wearing armor made out of crocodile or hippopotamus hide, or warriors from the East wearing suits of scale armor that made them look like metallic pine-cones. Sweat was one of the greatest enemies on campaign as this could rust armor in a day or disintegrate leather straps so that plates just fell off. This perhaps explains why Roman soldiers preferred to fight with bare legs, thus reducing the build up of heat beneath their body armor.

Recreating the experience of fighting in Roman armor is difficult, claims Graham Sumner, a member of the award-winning *Ermine Street Guard* re-enactment group: "The noise is something we have recreated. Numerous legionaries marching in full armor create such a noise that we have concluded that visual and not verbal commands would have been used, with musicians and standard bearers being used to convey these to ranks of soldiers virtually deafened by the sound of their own clashing arms and armor. The heat and exhaustion of moving in armor on a warm day is something we have also felt."

COH·I·BATAVOR

Roman Camel Soldier, Palmyra, 2nd Century AD

[Plate 4]

Palmyra was on the front-line of two great ancient empires, that of the Romans to the west and the Persians to the east. It grew rich on the trade routes that crossed the Syrian plain to the Persian Gulf and the Mediterranean. In the 3rd century AD, it was the center of a most remarkable realm lead by one of the very few recorded incidences of a female warlord.

Zenobia was the wife of Odaenthus, king of Palmyra. She was no average woman as the Roman chronicle *Historia Augusta* recalled: "Sometimes she walked with her foot-soldiers for several miles. She hunted with the eagerness of a Spaniard, often drank with her generals and sometimes drank with Persians and Armenians to get the better of them." She learned much from her husband. Odaenthus took political advantage when the Roman Emperor Valerian was captured by the Persian ruler Shapur I. He assumed control of the eastern Roman provinces and fought off the Persians. A grateful Roman empire acknowledged his governorship of all its eastern provinces. But in 267 AD Odaenthus was assassinated. Zenobia took over and was immediately sucked into a conspiracy against the Roman Empire making Egypt an independent realm.

Victory in Egypt encouraged Zenobia to march northwards and take on the Romans in Asia Minor. She extended her rule as far as the Bosporus, just across the water from Constantinople, capital of the eastern Empire. But then came a new Roman Emperor—Aurelian. He could not tolerate the defiant rule of Zenobia. The showdown came outside the desert town of Emesa, now called Homs. Clad in mail armor, Zenobia personally led her army, a mixture of heavy cavalry called *clibanarii*, armored head to toe, and light Arab horse-archers. Aurelian brought with him a variety of Roman soldiers, recruited from throughout the Empire, including Celts, but he also had warriors more suited to the local terrain, such as Moorish cavalry and perhaps a camel corps.

Knowing that Zenobia's strength lay in her cavalry, Aurelian tried to exhaust them by withdrawing before their charge. But this is a difficult maneuver to carry out successfully and the Roman cavalry panicked, breaking before Zenobia's horsemen. Zenobia spurred her own horse forward, but she had fatally dismissed the presence of the Roman legionary foot-soldiers. They held their formations and advanced forward, methodically dispatching the gangs of cavalry they surrounded. Zenobia fled but was later captured by Aurelian and led through the streets of Rome on a golden chain.

The painting opposite, by Graham Sumner, reconstructs a *Dromedarius*, a Roman camel-mounted soldier. He wears a scale armor shirt and a Roman bronze helmet based on one found in Syria. He wears a long tunic to protect himself against the heat and sand of the desert, with a scarf wrapped around his face. He is armed with throwing javelins, a bow, sword and long lance.

The Romans first encountered camel-mounted troops when they clashed with Seleucid king Antiochus III in 190 BC. At the battle of Nisibis, the Romans discovered that the soft feet of the camels were particularly vulnerable to spiked caltrops thrown in their way. The Romans waited some time before raising their own unit of camel soldiers—for the most part camels being used as important transport animals.

It was in the 2nd century AD that the first instance of a Roman camel corps was recorded. The Emperor Trajan had added a new Roman Province of Arabia Petraea (the region around Petra) to the Empire and a unit of Palmyrene camels had proved particularly effective for him so that he incorporated them into his army as *Ala I Ulpia Dromedariorum Palmyrenorium*. The fact that these camel warriors were Palmyrene in origin might mean that by the time Zenobia rose up against the Romans, the unit had changed sides and fought for Zenobia not the Romans. It has been estimated that the Romans had up to a thousand of these camel-mounted soldiers. For much of the time such warriors were used as policemen, patrolling the frontiers of the Roman realm and battling with desert bandits.

Cuchulainn, Mythic Hero, Northern Ireland

[Plate 5]

Ulster Warlord Cuchulainn is featured in the epic Irish tale Tain bo Cuailnge. Dated to the 8th century AD, the tale is believed to be centuries older, referring to the mysterious pre-Christian Celtic past of the island. It is a terrifying vision of the power of violence to transform a man into a monster.

"Cuchulainn began to work himself up into a fighting fury," wrote the chronicler, "a spasm tore though his body. It distorted him—made him a monstrous thing. Every bone and organ shook like a tree in a storm. His insides made a twist within his skin. His shins filled with the bulging muscles of his calves. Balls of sinew as big as a warrior's fist pumped up his body. His head swelled and throbbed." With his physical transformation completed, "dazzling lights and thick black smoke rose above his head." Cuchulainn was ready for battle and jumped into a blade-covered chariot.

The *Tain* is a brilliant evocation of ancient Irish warfare. It gives full weight to the importance of charisma in military leadership and describes the very real physical transformation attained by supreme fighting men. The military conflict described in the tale is far from epic, however, it being a cattle raid. But to dismiss this as dramatic exaggeration is to miss the point. Cattle raids were the very essence of ancient Celtic warfare—aside from the material gain, it was an opportunity to test the warriors of a tribe, introduce young men to the ways of violence, and to ensure that a tribe was respected throughout the region. It kept your neighbors on the back foot and promoted your own interests. It is interesting to note that chariots formed an important part of this type of warfare.

The painting opposite, by Angus McBride, is a reconstruction of Cuchulainn based on literary descriptions within the *Tain*. According to the tale, the warlord carried on his back a "dark-red curved shield", an "ivory-hilted sword" hanging from his waist, and in his hands he wielded a "five-pronged spear". The five-pronged spear was probably a reference to the legendary *gae bolga* or "Lightning spear" which entered a man as one blade and then exploded into a multitude of barbs ripping the victim apart.

Again, with reference to the *Tain*, Cuchulainn wears a thigh-length shirt or *lein* of "waxed skin, plated and pressed together, and fastened with strings and cords and straps against his clear skin." Over this he wears a "deep battle-belt of stiff, tough, tanned leather from the choicest parts of the hides of seven yearlings." The belt is part of his armor "to repel spears or spikes, javelins, lances or arrows—they fell from it as though dashed at stone or horn or hard rock." In addition to this, he wore a "silk-smooth apron, with its slight-gold speckled border, up to the softness of his belly. Over this silky skin-like apron he put on a dark apron of well-softened black leather from the choicest parts of the hides of four yearlings, with a battle-belt of cowhide to hold it."

Not shown in the painting but apparently typical of ancient Irish warlords was the crested battle-helmet. Ferdia, Cuchulainn's adversary, is described wearing a "war-like crested battle-helmet, finely decorated with forty precious carbuncles and inlaid with red enamel and crystal and gleaming stones from the East." Is this a reference to a Roman or Byzantine helmet that somehow made its way to Ireland? In a later version of a tale from the Finn cycle, there are further references to the special helmet worn by Irish warlords: "to protect the hero's head in battle he seized his crested, plated, four-edged helmet (*cathbharr*) of beautiful refined gold with bright, magnificent, crystal gems and with flashing, full-beautiful, precious stones which had been set in it by the hands of master-smiths and great artists."

Central Asian Warriors, Tajikistan, 7th Century AD

[Plate 6]

Throughout the ancient world, the great civilizations, such as Rome, Byzantium, or China, have been troubled by raiders from the wilderness. The Romans called these raiding tribesmen "barbarians", a word denoting all people who lived outside of the civilized world. The most fierce of these barbarian peoples were those tribesmen who came from Central Asia, from the steppes of southern Russia and Siberia. They were superb horsemen and excellent archers. The ancient Greeks first called these warrior raiders "Scythians". The Romans later knew them as "Huns".

Other later waves of tribesmen led similarly successful raiding campaigns that took them deep inside imperial boundaries. Some of the most famous of these were the Avars, the Mongols, and both the Seljuk and Ottoman Turks. Today, the descendents of these ferocious raiders still live in Central Asia. We now call these peoples "Turkic". Many of them are still involved in battles with imperial powers, such as the Chechens in their conflict with Russia. It is and will ever be a never-ending conflict between raiders and civilization.

Russian archaeologists discovered depictions of some of these ancient Turkic warriors in the wine-producing town of Pendzhikent in Tajikistan near the border with Uzbekistan. The town prospered in the 7th century just before the Arab conquests made this land part of the Islamic world, where it remains today. The warriors who rode out from Pendzhikent were the descendants of the White and Red Huns, who raided and conquered the Sassanian Persian Empire—they are also known as Hephthalites. The so-called Black Huns were led by Attila and turned west to raid the Roman Empire.

The Hephthalite Huns were a ferocious horse people. They devised new forms of saddle and pioneered the use of the stirrup. But they were also tremendous merchants and the region grew rich on trade passing along the Silk route from the west to China. It was no doubt the great wealth that attracted the Arab conquerors to this land. At first, the Turkic horse warriors bitterly resisted the Arabs, but the merchants of their towns could see the sense of becoming part of a larger trading zone which reached deep into the Middle East and came to a deal with the Muslims. In 709 AD Bukhara and Samarkand became part of the Umayyad caliphate in Damascus. Their Turkic rulers were left in place. But subsequent rebellions against Arab rule were crushed.

The painting opposite, by Richard Hook, reconstructs two Turkic horse warriors based on wall paintings found at Pendzhikent. The warrior on the right is a member of the ruling elite and is fully armored in a combination of mail shirt and mail aventail, over which is strapped a bronze lamellar cuirass. His arms and legs are protected by hoops of bronze armor. The two feathers worn on his segmented iron helmet indicate that he is an expert archer, able to shoot from horseback both forwards and backwards. He is heavily armed with a sword, dagger, bow and lance.

The mounted warrior on the left is a member of the leading warrior's retinue. He wears a long mail hauberk with further pieces of mail covering his legs. He wears a simple padded cap that would usually be worn under an iron helmet. His straight sword is modeled on western types of blade. Note that both warriors use stirrups to ride with.

The armies of Central Asian Turkic warlords would mainly have been composed of cavalry, though foot-soldiers could be raised from the local agricultural workers. The widespread use of mail suggests considerable wealth in the region and it is likely that for both cultural and economic reasons, most warriors entered battle on horseback. Horse-tail banners were typical standards carried in combat—as were windsock style banners with gaping mouths shaped like wolves, dragons or fighting camels. The Arabs included Turkic horse-archers in their armies after their conquest of Central Asia.

Viking Clashes with Irish Warlord, Southern Ireland, 1014

[Plate 7]

The battle of Clontarf in 1014, fought just outside Dublin, is a famous Irish victory over the Viking invaders, celebrated especially as most of England had already submitted to the Norse conquerors. In reality, it was more a struggle for power between two Irish factions disputing the High Kingship of all Ireland, with the Vikings playing opportunistic politics, rather than a concerted effort to rid the country of the Viking settlers. And yet, the achievement was significant and rang loudly through the chronicles of the period. The combat between the Irish warlord Murchad, son of the legendary Brian Boru, and the Viking chieftain Sigurd, was considered the highpoint of the battle and is recreated in the painting opposite by Angus McBride.

The combat between the two warlords was a classic heroic duel typical of early medieval Celtic literature. Whether such a clash actually happened is impossible to tell, but there is a strong tradition in medieval warfare in which leadership depended on being seen at the center of the action and a noble reputation could only be won by single combat with a warrior of similar high rank. On such occasions, noble warriors singled out their opponents and engaged in combat while more lowly warriors held back to watch the outcome (the same happened in the Far East in samurai warfare). Murchad was just such an ambitious warrior and would no doubt have been keen to add to his reputation, while his father, Brian Boru, sat at the rear as commander-in-chief behind a shield wall. In the end, after a long and exhausting duel, Murchad won the combat and cut the head off the Viking, but the next day Murchad died of his wounds.

Contemporary Irish chroniclers of the battle of Clontarf characterized the Vikings as cunning, professional warriors wearing coats of triple-layered mail, whereas Brian Boru's men were quick and bold and wore only handsome, multi-colored shirts. Clearly, this was an heroic convention in which the wearing of armor was seen as something devious, while the sons of Brian Boru were all the braver and worthy for entering battle unarmored—and yet still they defeated their iron clad enemy. In truth, the leading warriors of Brian Boru's army would have had as much access to mail armor and the finest weapons as the Viking settlers did. Indeed, they probably bought or captured such equipment from the Norse in the first place.

Murchad, the warrior on the left, is portrayed wearing a shirt of mail and wielding two swords similar to those found in archaeological sites around Dublin and similar to those used by the Vikings. That said, his headgear shows a distinctly Irish preference for heroic garb over the latest and most effective armor. Murchad wears a crested helmet studded with semi-precious stones, as described in the *War of the Gaedhil with the Gael*: "They had on them also crested golden helmets set with sparkling transparent brilliant gems and precious stones on the heads of chiefs and royal knights." Though no such helmets have been found, the *cathbharr* seems part of a specifically Gaelic tradition of noble crested helmets that goes back to the northern Irish myth-cycle of the *Tain*.

The Vikings first came to Ireland in the 9th century. The *Annals of Ulster* records them as raiders: burning, slaughtering and plundering. "They made spoil-land and sword-land," recorded one Irish chronicler, "they ravaged Ireland's churches and sanctuaries and destroyed her reliquaries and books. They killed Ireland's kings and chieftains and champion warriors. They enslaved our blooming, lively women, taking them over the broad green sea."

The Vikings built fortified towns at river mouths and these developed into the towns of Dublin, Waterford, Wexford, Cork and Limerick. But the Irish fought back. Their principal leader was the warlord Brian Boru. He launched relentless raids on the Viking settlements. At the battle of Clontarf he defeated the Vikings and their Irish allies who wanted to challenge his lordship, but in the last moments of the fight he was cut down by a Viking. The Vikings stayed in Ireland but they never dominated the Irish

Viking Warlord, Norway, early 11th Century

[Plate 8]

The Vikings are among the most notorious of early medieval warriors. They are the northern European equivalent of the Huns or the Mongols—famous for the violence and savagery of their raiding. They came at a time when Europe believed the age of the barbarians was over. It was a cruel shock to the early Christian world.

In 793 AD, their arrival in Britain was recorded in the Laud Chronicle, accompanied by strange events: "In this year, terrible portents appeared over Northumbria and miserably frightened the inhabitants. There were flashes of lightning and fire-breathing dragons were seen flying in the air. A great famine followed these signs and a little after that raiding heathens destroyed God's church in Lindisfarne by rapine and slaughter."

By the 11th century, however, many of the Vikings had been converted to Christianity. King Olaf Haraldsson was a prominent champion of Christianity in Norway. He is also known as St Olaf and is the patron saint of his country. In 1016, he took control of Norway and set about securing his control over local warlords by forcing them to give up their old pagan ways and embrace the new religion. Any who refused were tortured or killed and their property confiscated. It was a ruthless way of bringing his countrymen into the new faith.

On one occasion in 1018, Olaf was informed of a gathering of five pagan warlords near the Swedish border in Upland. They were plotting against him and raising a force of local warriors. The Norwegian king chose 400 of his own elite soldiers and acted quickly and decisively. The painting opposite, by Richard Hook, recreates the moment when Olaf crashed into the meeting of the five kings. Three were instantly banished and two others were mutilated as a warning to the rest.

King Olaf is shown wearing a Gjermundbu-style helmet, the only authentic Viking helmet that has been discovered. He wears a fine coat of mail plus ornate crucifix, sword, and dagger. A Viking chronicle describes Olaf wearing the following impressive garb: "Olaf wore a coat of mail, and had a gilded helmet on his head. He was girded with a sword with pommel and guard embossed with gold, and in his hand he held a barbed spear, chased and beautifully engraved. Before him he carried a red shield on which a lion was chased in gold." In the painting, Olaf's followers wear the plainer Spangenhelm typical of the period and carry the characteristic broad-bladed battleaxe.

Outside of Norway, the Vikings first operated as brutally effective raiders, seizing treasure and women and men as slaves to be taken back to their homelands. Their ships would land on a promising foreign beach and with either horses brought with them or stolen from farmers, they would ride deep into the countryside, until they had gathered more goods than they could carry. Their captured wealth meant they were equipped with the very finest of armor and carried excellent weapons. Mail tunics were worn by most of them and they carried large round shields reinforced with metal bosses.

But according to Viking sagas, such armor did not always protect its warriors: "Hrolf struck at Orn, but as he tried to fend off the blow with his shield, the sword sliced through and the point ripped open the whole of his belly so his guts poured out... Hrolf turned on Ulf and stabbed him with his sword. The mail coat failed and the sword went right through him. (*Saga of Gongu-Hrolf*)"

Having proved their superiority over Saxon defense forces in Britain, Viking leaders considered this a land suitable for settlement and in the 9th century major incursions were undertaken. Norwegian Vikings tended to settle in the north of the country and York was a great Viking settlement. The Vikings of Norway were the furthest traveled of all the Scandinavian raiders of the 9th and 10th centuries. Whereas the Danes hopped across the North Sea to eastern Britain, Norse ships took them across the ocean to the islands of Scotland and to Ireland. Further north, the Norwegians sailed to Iceland and Greenland and then onto North America, which they called Vinland.

Lithuanian Warriors, Prussia, late 13th Century

[Plate 9]

The Teutonic Order had begun as a Holy Order of Crusaders helping defend Christian conquests in the Holy Land. But when they shifted their base to northern Europe, they had to find a new enemy. They found them in the pagan tribes of eastern Poland and Lithuania. From their castles and towns in Prussia and along the Baltic coast, the Teutonic Knights led murderous campaigns against these peoples.

Ostensibly intended to convert the people they fought against, the campaigns deteriorated into bloodthirsty looting expeditions. It was treated more as sport and indeed, the Teutonic Order invited many knights from other European countries to join them on their pagan-hunting expeditions. The cynical truth behind these annual raids was revealed by the late 14th century when they continued despite the fact that Lithuania had converted to Christianity.

But the violence cut both ways. Lithuanian raiders turned on their Teutonic tormentors in major raids. In the winter of 1270, a plundering force of Lithuanians crossed Latvia and then used the frozen sea-ice of the Gulf of Riga to penetrate the Baltic island of Osel. The Bishop of Osel rallied his local troops and was joined by a unit of Teutonic Knights. Confident of an easy victory over the more lightly armed Lithuanian raiders, the Bishop clashed with them on the frozen sea. But this time the raiders were led by the Grand Duke of Lithuania and stood their ground, with many seasoned warriors among their ranks.

The Lithuanians defended themselves against the knights by pulling their sledges into a temporary fort. The Teutonic Knights led the assault but were caught up in the Lithuanian defenses and were cut down. The Bishop and his other warriors, including many Danes, pushed on through the sledges and forced many of the raiders back across the ice, but others remained and threatened to surround the Teutonic Knights.

Eventually, the Bishop was wounded and was forced to withdraw his force. The pagans had triumphed and took away their booty across the bloodstained ice. A Livonian chronicle described the scene: "There was a wild hacking and hewing on both sides, Christian and heathen. And the blood of men from both armies spilled onto the ice. It was a fierce fight in which many noble men were struck down."

The painting opposite, by Richard Hook, recreates two Lithuanian warriors of this period. The horseman is an aristocratic warrior of this period, equipped with a mail shirt and carrying lance, spears and sword. His helmet is based on one found in Prussia. The Prussians, benefiting from the great wealth generated by the merchant towns of the Hanseatic League, had the wealth to make iron helmets and mail, whereas there is no evidence that the Lithuanians ever manufactured their own mail. The mail shirt and helmet were either bought from a Prussian merchant or stolen from a dead Teutonic warrior.

The warrior on foot is more typical of Lithuanian native warriors, adapted to the snowy conditions by wearing a long fur coat and using skis and a sledge. He wears an Asian-style helmet which is again based on one found in Prussia, but was probably manufactured in either Poland or Russia where Turkish or Tatar influence was more evident in the form of arms and armor. Characteristically, he wears a feather in his helmet typical of Turkic raiders and his principal weapon is a composite bow. The light mace in his belt is typical of examples found in the Baltic region. Most Lithuanian cavalry preferred throwing javelins rather than bows.

Most Lithuanian raids were usually carried out in summer when their lightly armored warriors could use the local swamps most effectively. They knew the winter was preferred by the Teutonic Knight as their heavily armored horsemen liked more solid frozen ground. Their raids were mainly aimed at seizing slaves, cattle and other movable booty—they had no interest in conquering or holding land.

Edward Bruce Battles Anglo-Irish Warriors, Ireland, 1318

[Plate 10]

Edward Bruce was the brother of the more famous Robert Bruce, the Scots warlord who vanquished the English at the battle of Bannockburn in 1314. Following this remarkable victory, Robert Bruce planned to get the English on the run by uniting all Celtic peoples against them. He sent a famous letter addressed to all the chieftains of Ireland in which he proclaimed: "We and our people and you and your people, free since ancient times, share the same national ancestry and are urged to come together more eagerly and joyfully in friendship by a common language and by common custom."

But such a noble ideal came to nothing as Edward Bruce sailed to Ireland with an army of Scots marauders, who left a trail of destruction and animosity across Irelands as they looted and ravaged the local population. So much so that an Irish chronicler called them "Scots foreigners less noble than our own foreigners (that is, the English)." The battle of Faughart in 1318 saw the Irish chieftains of Meath and Down unite with the Anglo-Irish against a man who was no longer seen as a liberator, but just another blood-thirsty adventurer.

The painting opposite, by Angus McBride, recreates the moment at the battle of Faughart where Edward Bruce was overwhelmed by Anglo-Irish troops. The Scots warlord who had proclaimed himself High King of All Ireland is clad in armor. Opposite him is a true Irish chieftain, holding a javelin ready to cast the final blow. According to John Barbour's 14th century poem *The Bruce*, Edward Bruce wore an ordinary suit of armor at Faughart, without heraldic emblems, in order not to attract the attention of the enemy.

The mounted Irish chieftain in the painting is a fascinating figure, as he represents the Gaelic tradition of aristocratic warriors and wears so-called primitive garb. A 14th century manuscript illumination in Jean Creton's *Account of Richard II's Irish Expedition of 1399*, depicts such a warrior clad in a long shirt of mail with a saffron colored cloak and conical iron helmet, but his legs and feet are bare and he rides without stirrups.

Jean Creton's accompanying text describes a scene in which: "I beheld Macmore and a body of the Irish, more than I can number, descend the mountain. He had a horse without housing or saddle, which was so fine and good that it had cost, they said, four hundred cows... In coming down it galloped so hard that in my opinion I never saw in all my life any other animal run with such speed as it did. In his hand he bore a great long dart, which he cast with much skill." Jena Creton was a French clerk traveling with Richard II to Ireland so this is a first-hand account.

Creton's observations backed up those of Giraldus Cambrensis made in his *Topography of Ireland* two hundred years earlier: "Likewise, in riding, they neither use saddles nor boots nor spurs, but only carry a rod in their hand, having a crook at the upper end, with which they both urge forward and guide their horses. They use reins which serve the purpose both of a bridle and a bit... moreover, they go to battle without armor, considering it a burden, and esteeming it brave and honorable to fight without it."

From repeated written and visual evidence, it appears that Irish warriors continually affected "primitive" manners as a way of defining their difference from the English invaders. Their taste for fighting without stirrups and walking barefoot may also have been a bravura display of Celtic manhood and ferocity. Irish mounted warriors seem to have drawn on a centuries old tradition of Celtic horsemanship in which spears were thrown over-arm rather than carried as lances beneath the arm.

Edward Bruce lost the battle of Faughart and lost his life. His head was cut off and sent to the English King, thus bringing an end to the extraordinary but brief ambition of Robert Bruce to unite all the Celtic peoples against the English.

Polish and Lithuanian Warriors, Poland, early 14th Century

[Plate 11]

Poland has always been on the frontline of East and West, squashed between the superpowers of Germany and Russia—but never more so than in the Middle Ages. Then, it saw itself facing marauding Teutonic Knights in the west and raiding Tatars in the east. In the 13th century, it faced an even greater danger—the Mongols.

In 1237, Batu—nephew of Genghis Khan—and his leading general Subedei, led an army of Mongols into Russia. They took apart any resistance, warlord by warlord. They left a power vacuum which would be filled by the Golden Horde for centuries afterwards, commanding respect as far west as Poland. But then seizing on an apparent lack of concerted western resistance, the Mongols pushed on further into eastern Europe. They advanced during the winter when the great rivers and swamps were frozen—no longer providing a natural barrier to invasion.

In 1241, the Mongols entered Poland. Near the town of Liegnitz, the forces of the west chose to make a stand. Duke Henry of Silesia and the Grand Master of the Teutonic Order gathered a major force of Poles and Germans. The Mongols set about testing their cohesion with the typical steppe tactic of withdrawing swiftly before them. The German knights took the bait and charged, but after a few miles, the Mongols halted and resumed their attack. The ragged formations of the western army were now easy meat for the Mongols. By the end of the day, the Mongols had collected at least nine sacks of ears sliced off the heads of the defeated German and Polish warriors. The road was open to Germany, but instead the Mongols rode south to ravage Hungary.

The western army that faced the Mongols did contain knights clad in western European style armor, but the Polish contingent possessed many elements that would have been more familiar to the Mongols—including horse-archers and horse warriors wearing Asiatic-style armor. This combination of military influences is clearly shown in the painting opposite, by Richard Hook, depicting Polish and Lithuanian warriors of the early 14th century.

The figure on horseback is a Polish nobleman from the Duchy of Mazovia. His distinct pointed helmet is of Russian origin but is worn over a western-style mail hood and tunic. His iron lamellar cuirass is strongly influenced by similar items worn by the Mongols. His shield has the characteristic shape of those made in Lithuania. His boots and horse equipment combine Mongol and Russian influences.

The southern Polish knight kneeling on the ground is a much more conventional western-style figure. He wears mail armor and an iron battle-hat. The Lithuanian foot soldier on the left wears an Asiatic style helmet of Russian origin. The heavy throwing javelins he carries are typical of those used by both Lithuanian warriors on foot and on horseback.

In the wake of the Mongol invasion, Polish warfare continued to be a mess of internecine combat in which the Polish king battled with independent duchies. The Teutonic Knights were happy to take whatever side suited their own interests. To the east was the expanding realm of the Grand Duchy of Lithuania. This was the only force in the region strong enough to challenge the dominance of the Mongol Golden Horde and it took over some Russian provinces. Later in the 14th century, Poland and Lithuania joined forces to confront the Teutonic Order and eventually broke its power at the famous battle of Tannenberg in 1410. Lithuanian warfare was more Turkic in its nature than that of the Poles, although javelin-throwing horsemen were more predominant than horse-archers.

Irish Chieftain, Ireland, 14th Century

[Plate 12]

The Normans first came to Ireland in the 12th century. They brought their same methodical and vicious form of warfare that had seen them conquer England in the previous century.

The medieval chronicler Giraldus Cambrensis wrote an account of the conquering Norman campaign and was struck by the primitive nature of Irish warfare: "They go to battle without armor, considering it a burden and esteeming it brave to fight without it. They are armed with three kinds of weapons—short spears, light darts, and heavy battle-axes of iron, exceedingly well wrought and tempered. These they borrowed from the Norwegians (Vikings). In striking with the battle-axe they use only one hand, instead of both. When all other weapons fail, they hurl stones against the enemy."

Despite the modernity of their arms, Giraldus could see that the Normans would have their work cut out in the different combat environment of Ireland: "The Normans may be very good soldiers in their own country, expert in the use of arms and armor in the French manner, but every one knows that differs from the way of war in Ireland and Wales. In France, war is carried out across open plains; here, in dense woods and mountainous terrain. In France, victories are won by close-fighting ranks; here, by the charges of lightly armed warriors. There, quarter is given and prisoners offered for ransom; here, heads are chopped off as trophies and no one escapes."

As we have seen in a previous appointing, even Irish chieftains seem to have adopted a primitive garb as part of their Gaelic military character. But there was another strand of Irish arms and armor that was influenced more strongly by that of the more militarily advanced Anglo-Normans, and yet still retained key Irish elements. The painting opposite, by Angus McBride, reconstructs just such an Irish warlord of the 14th century.

The *Triumphs of Turlough*, written in the middle of the 14th century, describes in detail the arming of an Irish noble warrior: "The first piece brought to him was a trusty well-made *acton* (*cotun*), dense, close-ridged, from his lower throat to the point of his rounded knees. Over this he was invested with a loose mailshirt (*luirech*) of hard glittering rings that made a rough surface, close of texture was it, with gilded border. Over his mail he drew tight a round-star-studded belt, in was hitched a long blue-gleaming dagger (*ennach*). hanging ready to hand. He set on him a strong-plated conical helmet; took to him a broad sword, deeply fluted, having golden cross-hilt and tracery embellished scabbard, which he girded to his side."

In 1397, a Spanish traveler, Count John de Perilhos, described the noble warriors of Ireland as: "armed with coats of mail—they wear them girded, and they have throat-pieces of mail and round helmets of iron, with swords and lances. The swords are like those of the Saracens, which we call Genoese—the pommel is like an extended hand, the blades are long and narrow."

There is little visual evidence of armor in the 14th century in Ireland, but literary descriptions find a close parallel with the arms and armor shown in Irish tomb sculpture of a century later, such as at Roscommon and Glinsk. In the painting, the Irish chieftain is shown wearing trews and ankle boots with spurs. Over this he wears a long *cotun* or *gambeson* of vertically fluted leather. His mail tunic is shorter than the *gambeson* with the throat and upper chest protected by a four-pointed piece of mail that rests on his shoulders and may have included a hood.

On his head, the Irish chieftain wears a *barbut* similar in style to the ancient Greek Corinthian helmet. It is a dramatic helmet unique to Ireland at this period and is again typical of the difference between Irish and English warfare. The best surviving example of this helmet was found at Lough Henney in County Down and is now in the Ulster Museum in Belfast—it has been dated to around 1380. Made of steel with reinforcing bands of bronze around the eye and nose-guards, it corresponds closely to contemporary representations of such helmets, especially the broad cheek-guards and long nasal.

Burgundian Knight Fights Ghent Citizen, Flanders, 1453

[Plate 13]

Flanders in the late Middle Ages was one of the wealthiest regions of Europe. Its towns produced fine cloth and other manufactured goods that were then sold all over Europe. Money gushed back into the hands of its craftsmen and merchants and the rest of Europe looked on enviously—especially its warlords. How could they get their hands on this money?

The Duke of Burgundy had a feudal claim to be the overlord of this area and saw it as an opportunity raise taxes on the hard-working bourgeoisie of these towns. But the Flemish were not interested in paying yet another layer of imposed tax. In 1453, the citizens of Ghent rose up against Duke Philip the Good of Burgundy. With their considerable wealth they prepared their defiance well—fortifying outlying towns and villages to protect their supply lines. The Duke of Burgundy took these positions one by one, aiming for the last remaining stronghold of Gavere, to the south of Ghent.

The siege of Gavere began on July 16th and it was followed by a ferocious bombardment from the Duke's impressive array of artillery. The commander of Gavere, Arnold van Speck, tried to negotiate with the Duke but he was in no mood to spare the rebels. That night, Speck and his comrades disguised themselves as Burgundians and rode to Ghent for help.

Encouraged by English mercenaries, an army of 30,000 rode out from Ghent to attack the Duke in his camp near Gavere. In the meantime the citizens of Gavere had surrendered and true to his word of no compromise, the Duke hanged all their leaders. Once the army of Ghent arrived on the scene, some of their English mercenary leaders changed sides. The rest of the Ghentish army formed up with field artillery on their flanks and a solid body of highly motivated infantry armed with pikes. Much of it was hidden in a woods. The Burgundian army was mainly a mercenary force but highly experienced.

In the first stage of the battle, the Ghentish troops seemed to weaken before the Burgundian onslaught of artillery and archery. Then one of the Ghentish wagons loaded with gunpowder caught fire. The Burgundian cavalry saw this as an opportunity to push forward their attack. The Duke rode with his horsemen. But in their excitement they blundered into a group of Ghentish infantry screened by the woods. The Duke and his heavily armored companions found themselves surrounded by pike wielding foot-soldiers. He broke out and made a run for it but was attacking by more infantry who wounded his horse.

Bertrandon de la Broquiere, one of the Duke's senior squires and flag-bearers, came to his rescue. Olivier de La Marche described what happened next: "Bertrandon, seeing the peril that lay around him, lowered his lance to which a pennon was attached, and digging his spurs into his horse attacked the Ghentish, shouting out in a loud voice—'Traitors! Traitors! Would you kill your Prince?'" Apparently, this stopped the attacking infantry briefly, by which time others came to the Duke's rescue and he got away.

The Ghentish army was defeated and fled back to their town. The Duke had had a very lucky escape. Bertrandon was rewarded with mastery of the village of Vieux-Chateau and its income. Although Burgundy had won this combat, he did not possess the will or desire to occupy the town of Ghent. Flanders remained a serious challenge to outside authority for centuries to come.

The painting opposite, by Graham Turner, shows the Burgundian knight Bertrandon de la Broquiere in combat with a Ghentish foot soldier. Bertrandon wears the very finest in Gothic-style plate armor, including the characteristic *salet* helmet. The Ghentish figure is a citizen soldier and well-equipped wearing mail and a padded jacket.

GRAHAM TURNER '97

Hussite War Wagon, Bohemia, 15th Century

[Plate 14]

The Hussites were followers of Jan Hus, a man dedicated to criticizing the corruption of the Pope and the Catholic church. He encouraged the singing of hymns throughout the church service (previously reserved for special occasion) and he sang them in Czech, the language of his people, not Latin. The Czech people resented the influence of the Germans on Bohemia and the cause of Hus coincided with a surge of Slavic nationalism. In 1415, Hus was invited to explain his beliefs in Switzerland before a Catholic conference but he was condemned and burnt at the stake as a heretic. It lit a fire in Bohemia that led to the Hussite rebellion.

Jan Zizka, a warlord who had served the Czech royal family, attended many of these Hussite congregations and when revolution broke out he became their military leader. He knew the Czech king was sympathetic to the cause of the Hussites so did not consider his action a betrayal. But the King was nervous and called in the support of his brother Sigismund, king of Hungary. The Czech King then died from a stoke and Sigismund saw the military operation against the Hussites as an opportunity to seize control of Bohemia. With the blessing of the Pope, the campaign became a crusade.

In his war against Sigismund and his Hungarian and German knights, Jan Zizka developed a ruthlessly effective form of warfare. Knowing he faced a force strong in cavalry, Zizka knew he had to create a defensive formation on the battlefield, around which the enemy cavalry could exhaust themselves before Zizka then led a counter-attack of his highly motivated troops. The model he chose for this new form of warfare was one common to Eastern Europe in which troop formations frequently had to protect themselves against swarms of raiding cavalry.

Zizka took ordinary wagons and reinforced them with wooden boards and iron strips. Some even had loopholes carved into them through which handgunners and archers could fire their weapons. In battle, these wagons were chained together to form a temporary fortress with pavises and stakes filling the gaps between them. Cavalry could not break this mobile fieldwork and suffered terribly from the hail of shot coming out of them. This combined with Zizka's effective use of field artillery and handguns created a devastating offensive ability that shattered more traditional medieval armies.

The painting opposite, by Richard Hook, recreates one of these Hussite war wagons. Inside can be seen Hussite troops armed with handguns and crossbows. Foot-soldiers armed with flails wait to launch a counter-attack when the charging Hungarian cavalry outside have had enough. The chalice and goose banner of the Taborites, the most radical of the Hussite fighting groups, flies above the wagon.

In addition to Zizka's battlefield tactics, the Hussites also believed they had God on their side and their fanatical belief proved decisive in many conflicts. In 1420 at Vitkov hill just outside Prague, Zizka and his Taborites entered battle singing hymns. Hussite priests strode into combat carrying religious banners bearing the Chalice of the Host. "You warriors of God and his Law," the Hussite warriors were told, "pray for God's help and believe in him, so that with him you will ever be victorious. You archers and lancers of knightly rank, pikemen and flailmen of the common people, keep you all in mind the generous Lord." They did just that and this fanatical horde broke the Germans and Hungarians before them. The defeat encouraged Sigismund to avoid a siege of Prague.

Bohemian noblemen thought it wise to calm Sigismund by proclaiming him king of their realm, but Zizka and his supporters could never accept this and fought on against the invaders. Eventually, the conflict deteriorated into a civil war between various Czech factions with Zizka marching against Prague. Zizka's war wagons proved irresistible in battle and he continued to dominate Bohemia until he died in late 1424. His followers carried on with his method warfare for another ten years until more conservative forces prevailed. When they rode into battle they carried a banner with Zizka portrayed riding on a white horse and holding his famous fist-shaped mace. Even in death, Zizka led his army to victory.

Christian Warlords, South-Eastern Europe, 15th Century

[Plate 15]

1453 was a year of disaster for Christianity in south-east Europe. Constantinople fell to the Ottoman Turks and the Eastern Roman Empire was at an end. The Balkans now stood on the front-line in the vicious battle between Islam and Christianity. Hungary, Moldavia and Transylvania formed the new bulwark of the west and its kings and warlords were forced into the role of crusaders.

Matthias Corvinus became king of Hungary in 1458. He was the son of the great warlord Janos Hunyadi. Hunyadi dominated the region for ten years from 1446. He was the most effective weapon the west had against the Turks and he defeated them in several major battles, significantly slowing their advance in south-east Europe. He won his military experience against the Hussites and incorporated their tactics of firearms and war wagons into his own successful method of war. Tragically for the west he died from a disease caught while campaigning.

Matthias Corvinus inherited the military and political skills of his father and when he became king of Hungary he remained the region's most powerful figure for several decades. He hired mercenary warriors called the Black Army. They were mostly former Hussite troops, alongside German heavy cavalry, Poles and Serbs, who wore blackened army. They helped him win his victories and having secured his frontier against the Turks, turned west to capture Vienna and much of Austria. His victories did not outlast his death in 1490 and by the middle of the next century Hungary had been absorbed into the Ottoman Empire.

Stefan the Great of Moldavia rose to power in 1457. After the assassination of his father he lived in exile in Transylvania and Wallachia. It was there he made contact with Dracula, the Wallachian warlord who has gone down in legend as a vampire, but who was in fact a highly successful fighter against the Turks. Dracula helped Stefan invade Moldavia and seize the throne in 1457. In his first years in power he played a difficult game, allying himself with the Turks. This brought him into conflict with Matthias Corvinus and Dracula. When the Hungarians invaded Moldavia, Stefan defeated them at the battle of Baia.

In the 1470s, relations soured between Stefan of Moldavia and the Turks. The Sultan demanded tribute from him and when he refused, the Turks invaded his country. He withdrew before their far larger force, but then defeated them at the battle of Podul Inalt in 1475. With this victory, he declared himself a crusader against Islam and an alliance with Hungary followed. Unfortunately, the following year, the Ottoman Sultan returned to Moldavia and this time Stefan was defeated. But he returned to the fray and took the war against the Turks, installing Dracula as ruler of Wallachia. More indecisive fighting continued until a peace was signed with the Turks in 1485. A final successful repulsion of a Polish invasion crowned the military success of Stefan and he died in 1504, after reigning for 47 years.

The painting opposite, by Angus McBride, portrays the warlords Matthias Corvinus (on the left) and Stefan the Great. The horseman in the background is a Moldavian frontier cavalryman. Matthias is shown wearing plate armor in the German style with a *salet* helmet. The only Eastern flavor in his equipment is his curved saber. Stefan is clad in the more Eastern style of a light cavalryman. He carries a bamboo shafted lance and a shield characteristic of Balkan raiding horsemen. The cavalryman is typical of the lightly armored horsemen who fought the frontier wars against the Turks, carrying light spears, bow and curved saber—the ancestor of the Hussar.

Japanese Siege, late 16th Century

[Plate 16]

Samurai dominated ancient warfare in Japan for almost a thousand years, but their reign of sword and bow was coming to an end by the 16th century. It was a time of change, of relentless civil wars in which everyone fought each other for power, lords against lords, clans against clans, even the religious orders of Buddhist monks put armies into the field to fight against the samurai. Into this chaotic mix of feuding came a new invention, a weapon that would transform Japanese warfare just as it had transformed warfare across the other side of the world. That new weapon was the gun.

The painting opposite, by Richard Hook, recreates a samurai siege of about 1590. The besieging army has begun its assault on the samurai fortress by erecting several lines of palisades. Samurai warriors on horseback ride inside this no-man's land to test the firepower of the garrison. Behind the palisade are ramparts dug out of earth reinforced by rice bales stuffed with sand and bundles of green bamboo. From these ramparts, soldiers armed with guns keep up a constant barrage against the defenders.

But like in 16th century Europe, the archer has not been completely forgotten and a unit of bowmen operate from behind portable shields. There are also more old-fashioned style siege machines. Two kinds of wheeled assault towers can be seen on the left, one being pulled by peasants. In the center can be seen the entrance to a mine which is being dug under the walls of the fortress. Barrels of gunpowder could be taken down these tunnels and then ignited, or simply the tunnel supports removed, to bring down sections of the fortress walls. On the right, an observation box is being raised by a pulley. Oddly, despite the presence of firearms, there are no cannons being used in the siege. These only came into use from 1614 onwards.

The gun first came to Japan in 1542, when a ship containing Portuguese merchants found itself stranded on a strange island. Wading through the surf onto the beach, the traders were surrounded by a new people they had never seen before. A Japanese eyewitness describes the new weapon held by the Portuguese: "In their hands they carried something two or three feet long, straight on the outside with a passage inside, and made of a heavy substance... At its side there is an aperture that is the passageway for fire... The explosion is like lightning and the report like thunder."

The weapon brought by the Portuguese was the gun known as the arquebus, a matchlock operated firearm lighter than the musket and not needing a forked rest to support it. Immediately, the Japanese understood the potential of this weapon and the local warlord instructed his sword-smiths to copy the Portuguese guns—one such metalworker is believed to have swapped his daughter for information on some of the more difficult aspects of gun manufacture! Within ten years, Japanese-made guns had made their appearance on the battlefield and one warlord wrote to his followers telling them: "decrease the numbers of spears and have your most capable men carry guns."

As in Europe, the appearance of guns was not welcomed by everyone. The martial arts in Japan were highly evolved and many warriors had spent years perfecting their skills in archery and using the spear and sword, and just as knights in the West believed the new weapon was beneath their dignity to use, being unworthy of a warrior who won his reputation in hand to hand combat, so the socially superior samurai deemed the gun a weapon worthy only of the lowly classes of soldiers known as *ashigaru*. It is these troops who fire the guns in the samurai siege opposite.

Japanese Warship, late 16th Century

[Plate 17]

The samurai are largely viewed as a land-based force, but when compelled to go to sea they could be equally deadly, using their ingenuity to build some remarkable craft.

The painting opposite, by Richard Hook, recreate a giant samurai battleship in combat in the late 16th century. Such ships resembled floating castles and were covered with wooden planking 10cm thick. They may have been reinforced with iron, but this was not thought to have taken the form of iron plates—the honor of creating the first iron-clad battleships goes to the Koreans (see below). These giant samurai ships carried a tremendous amount of firepower. Muskets and arrows were shot through loopholes in the sides. But any such encounter at sea would usually have to be settled by hand-to-hand fighting. Warriors at the prow of the ship wait with giant grappling hooks to throw at the enemy ship and bring it closer so they can board it.

One of the most famous series of sea battles fought by the samurai was during the Japanese invasion of Korea in the 1590s. Their opponent was the wily Korean Admiral Yi who had invented a form of iron-clad battleship called the Turtle-ship. He had already surprised the Japanese by winning his first major maritime clash against them at Hansan-do in 1592. In retaliation, the Japanese sent 42 warships from their base at Pusan to crush the upstart Yi, pausing at Angolpo, midway between Pusan and Hansan-do.

Among the 42 Japanese craft were two giant flagships, one of three-storeys and the other of two-storeys, massive oared battleships that towered over Yi's Turtle-ships. Refusing to be drawn into the open sea, the samurai warships remained in the shallow waters of the port and Yi was forced to change his tactics. He divided his fleet into a number of small assault parties headed by the Turtle-ships that were ordered to attack the Japanese fleet piece by piece.

Just as he had done at Hansan-do, Yi instructed his Turtle-ships to attack the biggest and most important of the Japanese craft. Protected by their iron armor, the Turtle-ships rowed quickly towards the massive flagships where Japanese soldiers stood on each level of the many wooden platforms that sprung above their hulls. The Turtle-ships rocked as their cannons poured fire into the sitting targets, ripping the wooden structures apart in great splinters. The Japanese returned fire with their muskets and cannons, but their bullets merely ricocheted off the curved iron sheets covering the Turtle-ships beneath them. Height was of no advantage here. Only courage kept the Japanese flagships in the fight.

Rather than fleeing the burning boats, samurai warriors rowed ashore under fire to bring back reinforcements, their bravery being admired even by the Koreans. In the densely packed bay, the fighting consisted of broadsides followed by ramming and clashing of boats in which soldiers sometimes jumped on to each other's craft to engage in hand to hand combat. None could land on the Turtle-ships that were covered with iron spikes and these dragon-prow craft simply nosed among the chaos, blasting away at the wooden structures either side of them. All day, the warriors fought.

Eventually, the iron-clad Turtle-ships proved too tough for the Japanese to crack and despite their suicidal courage, the Japanese had to flee ashore and leave their boats to burn and sink slowly into the waters of the port. Yi gave them room to escape and retreated to the open sea where he spent the night. He would not destroy all their boats, preferring the enemy to leave by sea rather than by land. "Many bandits escaped into the mountains," he reasoned, "where local inhabitants had taken refuge. If all of the Japanese ships were to be burned, the bandits would massacre Korean citizens in revenge."

The Japanese had suffered a defeat that would change the whole nature of their war in Korea. Six days later, their commander issued a general order forbidding all Japanese naval forces to fight at sea against the Korean navy.

Calvach O'Carroll, Irish Chieftain, late 16th Century

[Plate 18]

Since the invasion of Edward Bruce in the 14th century, the rule of the English had declined in Ireland until by 1454 the Pale (the area of direct English control) measured a frontier of barely 150 miles. Tribal warfare and the power of individual Irish warlords had undermined English rule of the country. Descended from Norman conquerors, most of the Anglo-Irish lords had gone native, speaking Gaelic and wearing Gaelic clothing.

The process of English reassertion of power in Ireland began under the Tudor dynasty and reached a crescendo under Elizabeth I. Defiance broke out into warfare and both sides behaved ruthlessly. In 1570, Humphrey Gilbert was appointed commander of the English army in Munster and he did not hold back when it came to making his view of the Irish clear: visitors to his camp were compelled to walk between two lines of severed Irish heads leading to his tent. Under Gilbert, Irish independence was crushed in Munster and the region turned into English plantations.

In Ulster, in northern Ireland, Hugh O'Neill led a major rebellion against the English in 1595 with an army 6000 strong. He struck fear into the heart of the Elizabethan regime by negotiating with England's arch enemy—Spain. Three years later, the English sent an army under Sir Henry Bagenal to quell the uprising. But it was badly equipped: "the want of the men's apparel is such that if they not be speedily relieved, many will march without shoes or stocking." It was suggested they buy Irish clothing, but the Lord Deputy of Ireland refused to put the Queen's money into the hands of rebellious Irishmen.

The unprepared English force would be badly punished for their lack of cohesion. O'Neill avoided a set-piece battle but chose guerrilla action near the Blackwater. He had prepared the fighting ground brilliantly: brushwood and undergrowth woven together to create fences; pits dug to hinder cavalry; boggy land linked by a trench protected by a thorny hedge. Together, this maze of death scattered the English forces and they were hunted down in sections.

Emboldened, Hugh O'Neill was joined by Spanish soldiers landing at Kinsale. But the English confronted him with more conventional pike and shot tactics. The Irish and Spanish failed to act together and the English triumphed. O'Neill retreated to his power base in Ulster. In the end, it was not force of arms but a harsh winter that finally compelled him to surrender.

The painting opposite, by Angus McBride, recreates the appearance of Calvach O'Carroll, an Irish warlord who succeeded to the Chieftaincy of Ely in 1582 and held it for 18 years, winning for himself a reputation as a "warlike and defending man, and a strong arm against his English and Irish neighbors."

In Fynes Moryson's *Itinerary* of 1617, Irish chieftains are said to wear the following: "The gentlemen, or lords of countries, wear close breeches and stockings of the same piece of cloth (trews), of red or such light color, and a loose coat, and a cloak or three-cornered mantle." McBride's portrait is based on two woodcuts by John Derricke from his *The Image of Irelande* of 1581. They show a chieftain preparing for a raid and dining in the open. With this in mind, Calvach is shown wearing a typical Irish cloak or mantle with a fringe of rough wool. This has been called a tufted mantle, the effect being achieved by sprinkling the drawn nap with honey or sugar and vinegar and then using a bag of pebbles to curl it. Calvach wears a particularly Irish form of helmet, called the *cathbharr*, an iron decorated helmet said to have been worn by noble Irish warriors throughout the Middle Ages.

Angus McBride

Irish Warriors at Court of Queen Elizabeth I, late 16th Century

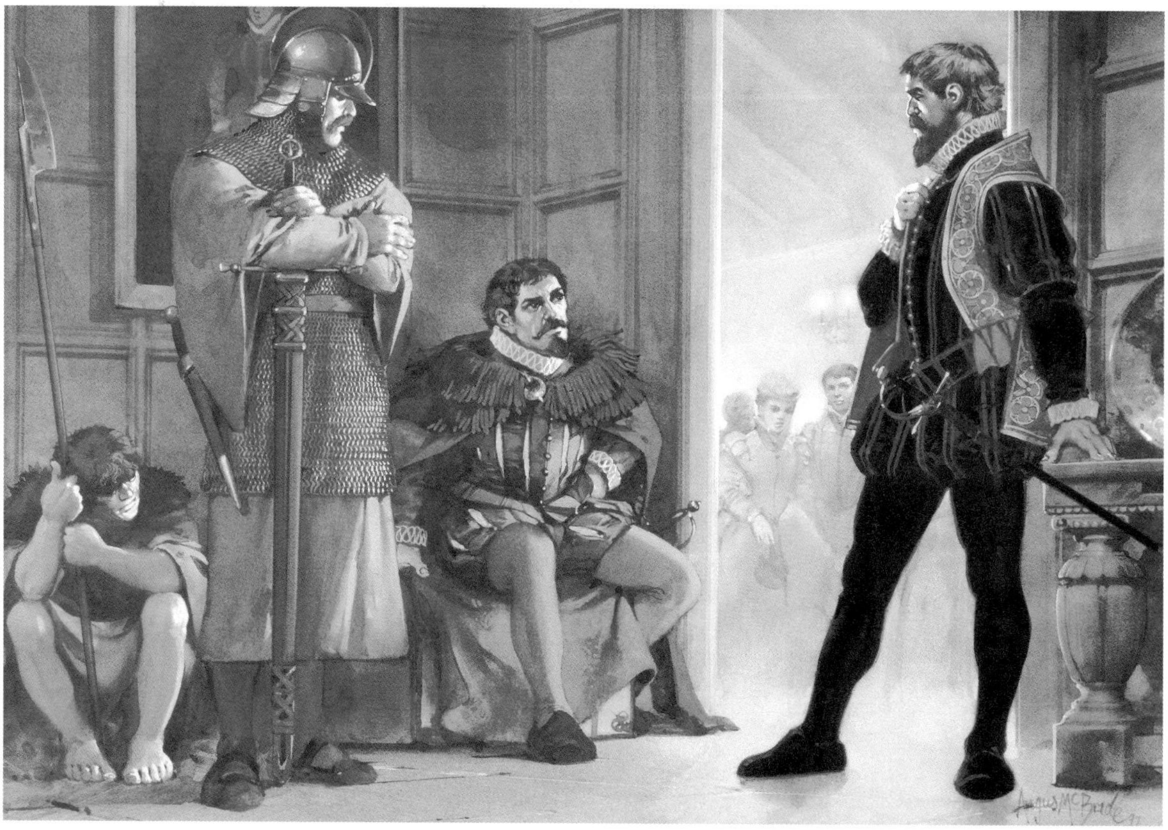

[Plate 19]

The painting opposite, by Angus McBride, is inspired by a famous scene described by the Tudor historian William Camden: "And now Shane O-Neill came from Ireland, to keep the promise he had made a year before, with an escort of galloglas, armed with battle-axes, bare-headed, with flowing curls, yellow shirts dyed with saffron, large sleeves, short tunics and rough cloaks, whom the English followed with as much wonderment as if they came from China or America."

This exotic description records the moment when Shane O'Neill—warlord of Ulster in northern Ireland—visited Queen Elizabeth I in London in 1562 to put his case for overlordship of Ulster. As a demonstration of his Gaelic claim to power, he and his followers wore traditional Irish clothes at the Tudor court. The painting opposite depicts a Scots galloglass and his Irish kern attendant awaiting their lord, Shane O'Neill, in a chamber of the royal palace at Whitehall.

To wear traditional Irish costume at the English court was very much a statement of political independence. Previous English laws had forbidden the wearing of Irish garb and declared the act tantamount to rebellion. In 1537, Henry VIII had passed an act banning the wearing of overly long and full garments as well as the use of saffron, a distinctly Gaelic dye. In defiance of this, Shane O'Neill's men wore "yellow shirts dyed with saffron and large sleeves." They even wore their hair long in a fringe or *glib*, despite English laws demanding it be cut.

Laoiseach Mac an Bhaird, a 16th century Irish poet, crystallized the importance of national appearance, especially hair style, in his lament for an Irishman adopting an English style: "O man who follow English ways, who cut your thick-clustering hair, graceful hand of my choice, you are not Donnchadh's good son. If you were, you would not give up your hair for an artificial English mode."

It is impossible to tell exactly where Shane O'Neill formally met Queen Elizabeth as they she had eight palaces in and around London, but it seems likely the meeting was held in Whitehall palace in central London. The best contemporary visual references for this remarkable display of Gaelic costume are Albrecht Durer's pen and wash drawing of 1521 depicting galloglas warriors and Lucas de Heere's water-colors of around 1575. The galloglas recreated opposite holds a two-handed sword and wears a coat of mail with a peaked steel helmet under which he wears a long tunic dyed with saffron: "They stain their large wide shorts with a saffron color... with the boughs, bark, and leaves of the poplar tree... adding thereto the rind of the wild arbutus tree, salt, together with saffron."

Galloglas were elite mercenaries employed by the Irish but recruited from the Scottish islands and western Highlands, being a fiery mix of Scots, Irish and Norse ancestry. Galloglas means "foreign young warrior" and some had fought with Edward Bruce when he invaded northern Ireland from Scotland 200 years earlier.

The young man crouching on the floor next to the galloglas, in the painting opposite, is a kern attendant. These were lightly armed warriors and he is shown wearing the characteristic *glib* hairstyle: the back of the hair is short, but the front fringe is long, hanging down to cover the eyes.

In battle, the kern won grudging praise from the English. "The kern is a kind of footman," wrote John Dymmok, "slightly armed with a sword, a target (shield) of wood, or a bow and sheaf of arrows with barbed heads, or else three darts (javelins), which they cast with wonderful facility and nearness, a weapon more noisome to the enemy, especially horsemen, than it is deadly; within these few years they have practiced the musket and caliver, and are grown good and ready shot."

Yermak Conquers Tatars, Siberia, 1581

[Plate 20]

As other European nations turned westwards to the New World to establish empires, Russia decided to turn eastwards towards Siberia. Constantinople had fallen to the Ottoman Turks in 1453. It was the "second Rome" to fall to barbarians from the east. By the beginning of the 16th century, Moscow rose as the Third Rome and its ruler, Ivan the Terrible, proclaimed himself Tsar—another name for Caesar—of the new Russian Empire. From that point onwards, the Russians threw off the Tatar yoke and expanded their eastern frontiers across the territory of the Tatars, Mongols and Turks of central Asia.

In 1574, Moscow instructed the Stroganov merchant family to establish trading strongholds on the eastern slopes of the Ural mountains—in effect an invasion of Siberian Tatary and a direct challenge to the warlord Kuchum. The Stroganovs were hugely wealthy, dealing in most raw materials, including furs, salt, iron and copper, and were the principle engine behind much Russian expansion. They took their time, however, no doubt preferring less hazardous tasks, but when they moved in 1581, they hired a Volga Cossack, an *ataman*, a Cossack leader called Yermak.

Cossacks in the 16th century were largely derived from the Turkic population they lived among. The word *kazak* means free man and they operated as freelance agents of the government, receiving land and tax remissions in return for guarding the frontiers against the Tatars. The Cossack method of fighting was typical of all steppe warriors, light horsemen armed with bows, spears and swords, frequently preferring ambushes and night attacks to open confrontation. They quickly adopted firearms in the 16th century that gave them a definite advantage over the bow wielding Tatars.

Being based on the great Russian rivers, many Cossack groups were also experienced sailors. Yermak Timofeyich was just such a warrior boatman and may well have worked in the fleet of the Stroganov family on the Volga. One chronicle says he left his work to gather his own small band of boatmen and together they acted as pirates with Yermak as their leader.

The size of the army Yermak put together to invade Siberia is also subject to debate, but two chronicles say the band came from the Volga and numbered 540 men. A further 300 men were then employed on their way to Siberia, making a total of 840. Such figures give the impression of a conquistador-style group of adventurers.

The adventure began on the river Kama and followed the rivers eastwards through the low passes of the Urals: Kama, Chusovaya, Serebryanka, land crossing to Zharavlya, Tagil, Tura, Tobol, Irtysh. No doubt the main mode of transport were the hollowed out log boats typical of the Cossacks. According to the Stroganov chronicle, Yermak's army included, in addition to the core force of Volga Cossacks, some Lithuanians, Tatars, Russians, and even some Germans or western Europeans. They were paid a satisfactory reward, attired in new clothing and given "firearms, cannon, and rapid-firing guns and seven-span muskets and many provisions and all these things in sufficiency." Together with interpreters of the Tatar language, they set off in September 1581.

Building a stronghold fortified with walls of earth on the Ural frontier on the river Serebryanka, Yermak and his Cossack conquistadors entered Siberia and the khanate of Kuchum when they passed on to the river Tura that took them down the eastern flank of the mountain range. They captured several Tatars and sent one of them to inform Kuchum of their arrival.

"The Russian troops are so powerful," the Tatar told the Khan. "When they shoot from their bows, there is a flash of fire and a great smoke issues and there is a loud report like thunder in the sky. One does not see the arrows coming out of them. They inflict wounds and injure fatally, and it is impossible to hide oneself from them by any trappings of war. Our scale armor, armor of plates and rings, cuirasses and mail do not hold them, they pierce all of them right through."

The painting opposite, by Richard Hook, reconstructs the clash between Yermak and his Cossack conquistadors and the Tatars during his expedition into Siberia in 1581. Yermak stands in the boat firing his pistol. Behind him stands a man holding the banner of St. Michael, underlying the crusading aspect of the expedition. The Tatars did not have guns, but fight back with bows and arrows.

Micronesian Warrior, Gilbert Islands, late 19th Century

[Plate 21]

Before the arrival of Europeans in the 19th century, warfare was virtually unknown among the native Gilbert Islanders in the south Pacific. In fact, it was impossible to fight a war in a European sense because every islander regarded fighting for others as a form of slavery and no one accepted anyone else as superior to themselves. There were no chiefs and the name of one of the larger islands, Tabiteuea, means "kings are forbidden." Everyone believed he was a "king" in his own right and no one was a servant or a slave.

As a result, fighting took the form of personal disagreements over land ownership or jealousy over wives and mistresses. Killing a person was a crime and could result in one's own death at the hands of avenging relatives unless major compensation was received. The art of fighting therefore was to inflict enough pain on your enemy without causing him death—the result was the development of elaborate armor.

The training of a Gilbert Islander in the art of fighting began at the age of 12 when his hair was cut painfully and slowly with a shark's tooth. This was accompanied by incantations that would give him a true warrior's heart and protect him against the attractions of the opposite sex. At the age of 20, his hair was cut again, but this time a torch was lit over his head with the sparks allowed to fall on his naked skin. The young man should not flinch or else the whole ceremony would have to be repeated.

Later, he would be given a drink in a coconut shell made of seawater and oil mixed with a stingray's barb. He would then be expected to sit still from morning to night while his father stabbed his head with a shark's tooth until the blood streamed down over his eyes and his cheeks. Any movement would be punished with a whipping blow from a palm frond. Having inured himself to pain, the young man was now ready to take on the weapons and armor of a fully-fledged Gilbert Islands warrior.

Gilbert Island weapons were simple but intended to strike fear and cause maximum pain. They adapted the most vicious elements of the natural world around them and strapped them to swords and spears. The serrated teeth of sharks were studded into swords two and half feet long and spears of seasoned coconut wood twelve or more feet long. The weapons were used in a slashing motion with the intention of tearing the flesh and leaving fragments of the sharks teeth in the wound to cause infection.

A similar concept lay behind the making of a spear four to five feet long to which the poisonous barbs of the stingray fish were attached, the barbs being burned slightly so as to be more likely to break off in a wound. This was biological warfare at its most primitive. To ward off blows from these swords and spears, three or four pieces of wood were attached to spears to form a trident that might trap the opponent's blade.

Even more impressive than their shark tooth weapons were the suits of armor worn by the wealthiest Gilbert Islanders. These were made entirely of coconut fiber that was woven into a thick mail that protected their legs, arms and body. To this might be added a breastplate of strips of dried ray-fish skin. On their heads they wore coconut fiber skullcaps over which were placed helmets made out of the dried spiky skins of inflated puffer or porcupine fish. To the back of the body armor might be attached a tall shield of coconut fiber which rose over the head to protect the back from stones or other missiles, frequently thrown by the women of the combatants if it was a single duel between two men. Equipped like this, the Gilbert Island 'knights' were ready to take on anyone, including the Europeans that appeared over their horizons in the early 19th century.

The painting opposite, by Richard Hook, recreates a Micronesian warrior from the Gilbert Islands wearing the full panoply of coconut fiber armor, spiky puffer fish helmet, and sharks teeth studded spear.

Mongol Warrior, Mongolian People's Army, 1921

[Plate 22]

When an empire collapses, chaos and barbarity is unleashed. It is a time for a man from nowhere to ride into that chaos and take control of it, creating for himself a new empire. It was this barbarian vision that captured the imagination of Baron Ungern-Sternberg in the second decade of the 20th century—for Baron Ungern-Sternberg knew he was the reincarnation of Genghis Khan, the greatest barbarian warlord of all time.

Roman Nicolaus Fyodorovich von Ungern-Sternberg came from a Russian family living in the Baltic state of Latvia. A violent spirit of enterprise ran in his veins for his forefathers had fought as mercenaries, crusaders and pirates. In 1908, he became an officer in a Cossack regiment and served in the desolate land of Central Asia on the border between Mongolia and China. His hungry mind was filled with stories of Genghis Khan and Attila the Hun and this was translated into reality when he was put in charge of a troop of Mongol horsemen.

Ungern-Sternberg immediately found a kinship with them, admiring their quiet acceptance of the harshest circumstances, their innate martial skills, and their fearlessness in battle. He wanted to be like these silent, frightening warriors and made a step towards this by adopting their religion of Lamaistic Buddhism. This gave him the mental strength he needed and it gave him a taste for the supernatural and the occult. He now believed in reincarnation and strange thoughts began to preoccupy him.

In 1917, the Russian Empire was on its knees in its war against Germany and a small group of Bolshevik revolutionaries took advantage of the political turmoil to lead a *coup d'etat*, taking control of government and setting about establishing a Communist regime. In the confusion that followed, the Russian Empire collapsed and all its realms were up for grabs. It was the barbarian chaos that Baron Ungern-Sternberg had dreamed of and with it came great opportunities.

By 1920, the great civil war between the White and Red Russians had ground to a halt with the crushing of the counter-revolutionaries by the Bolsheviks. Baron Ungern-Sternberg found himself out on a limb in the middle of Siberia. With his beloved Czar murdered and the White Russian army disintegrating, he now sought his own barbarian destiny. The land of the Mongols he so admired had been invaded by the Chinese, taking advantage of Russia's misfortunes to increase its own empire.

As the reincarnation of Genghis Khan, the mighty Mongol warlord, Ungern-Sternberg believed it was his duty to release the Mongols from their Chinese yoke. With an army of ex-White Army Russians and native warriors, he set out to rebuild the Empire of the Mongols. Ungern-Sternberg advanced into Mongolia in October 1920. The target of Ungern-Sternberg's campaign was Urga, now Ulan-Bator, the capital city of Mongolia. On October 26th, Baron Ungern-Sternberg led his first attack on the Mongolian capital at night. Despite an initial formidable onslaught, Ungern-Sternberg's warriors were outnumbered by the Chinese and well-entrenched machinegun crews caught them in a savage crossfire. Swept by snowstorms, the Baron retreated to the hills and waited five days. He then attacked for a second time, but was again beaten back.

With the severe Mongolian winter descending, Ungern-Sternberg retreated east and then south, placing himself across the crucial supply road from Peking to Urga. Living by looting the surrounding countryside, he embarked on a vicious cat-and-mouse campaign. Hounded by the Red Army and Mongolian Communists, the White horde eventually disintegrated.

A few die-hards accompanied the Baron into the wilderness, but he was badly wounded. Bleeding, exhausted and helpless, he slid from his saddle and fell to the ground unconscious. Left by himself on the steppe, a group of Mongols happened upon him. Unable to offer any resistance to the Mongols, the Baron just sat awaiting his fate, but the bandits were too afraid to kill him. If he was, indeed, a reincarnation of the mighty Khan then that would bring profound bad luck upon themselves. But they could not let him go and rekindle the civil war that had devastated their homeland. Their solution was to bind him with ropes and leave him to be eaten by the giant ants that covered every corpse in that part of the steppe.

The painting opposite, by Richard Hook, recreates a Mongol warrior of the Mongolian People's Army in 1921. Combining the power of modern weapons with traditional clothes and steppe pony, these were the ferocious warriors that hunted down Baron Ungern-Sternberg.